My Math Notebook

NAME:

www.ingramcontent.com/pod-product-compliance
Lightning Source LLC
Chambersburg PA
CBHW062329220526
45469CB00008B/2639
* 9 7 8 1 7 2 6 1 9 1 2 0 3 *